DEADLY EXPEDITIONS

THE GREELY EXPEDITION'S FATAL QUEST FOR FURTHEST NORTH

by Golriz Golkar • illustrated by Ana Carolina Tega

raintree
a Capstone company — publishers for children

Raintree is an imprint of Capstone Global Library Limited, a company incorporated in England and Wales having its registered office at 264 Banbury Road, Oxford, OX2 7DY – Registered company number: 6695582

www.raintree.co.uk
myorders@raintree.co.uk

Hardback edition text © Capstone Global Library Limited 2024
Paperback edition text © Capstone Global Library Limited 2025

The moral rights of the proprietor have been asserted. All rights reserved. No part of this publication may be reproduced in any form or by any means (including photocopying or storing it in any medium by electronic means and whether or not transiently or incidentally to some other use of this publication) without the written permission of the copyright owner, except in accordance with the provisions of the Copyright, Designs and Patents Act 1988 or under the terms of a licence issued by the Copyright Licensing Agency, 5th Floor, Shackleton House, 4 Battle Bridge Lane, London, SE1 2HX (www.cla.co.uk). Applications for the copyright owner's written permission should be addressed to the publisher.

British Library Cataloguing in Publication Data
A full catalogue record for this book is available from the British Library.

ISBN 978 1 3982 5137 3 (hardback)
ISBN 978 1 3982 5138 0 (paperback)

Editorial Credits
Editor: Abby Huff
Designer: Dina Her
Production: Tori Abraham
Originated by Capstone Global Library Ltd

All the internet addresses (URLs) given in this book were valid at the time of going to press. However, due to the dynamic nature of the internet, some addresses may have changed, or sites may have changed or ceased to exist since publication. While the author and publisher regret any inconvenience this may cause readers, no responsibility for any such changes can be accepted by either the author or the publisher.

Printed and bound in India.

CONTENTS

Exploring the unknown.................. 4
Scientific data and furthest north 7
Waiting for help 10
A dangerous journey 14
The beginning of the end 18

    Map of the expedition 28
    More about the expedition 29
    Glossary 30
    Find out more 31
    Author bio 32
    Illustrator bio.................... 32

Exploring the unknown

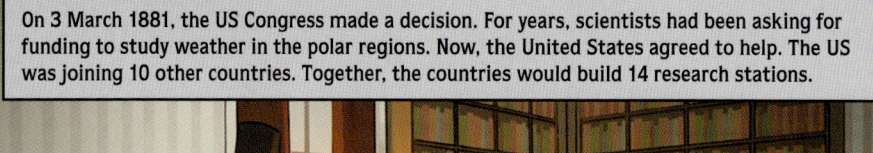

On 3 March 1881, the US Congress made a decision. For years, scientists had been asking for funding to study weather in the polar regions. Now, the United States agreed to help. The US was joining 10 other countries. Together, the countries would build 14 research stations.

It will be good for America!

It's a waste of money!

The US would build two stations. One would be in Alaska. The other would be near Lady Franklin Bay, about 600 miles from the North Pole. Some Congress members were against the expedition. Others saw an opportunity.

The Lady Franklin Bay team will have another, secret mission. To reach Furthest North!

Furthest North is the record for highest latitude reached by explorers. The British had held the record for centuries. The US hoped to break it.

After a smooth five-week journey, the crew arrived in Lady Franklin Bay. They stopped on the coast of Ellesmere Island. They unloaded 350 tonnes of food and supplies.

"Almost no ice in the waters! Most unusual for this time of year."

"Indeed, Lieutenant Lockwood! I feared ice might block our ship."

Unlike other polar expeditions, the crew would not keep their ship. It would soon leave. In a year, a ship would bring fresh supplies. In two years, a ship would bring the men home. Between those times, the men would have no connection to the outside world.

The crew built their headquarters, Fort Conger. They got ready to live and work in the harsh Arctic.

Waiting for help

As summer drew closer, the crew watched for the promised supply ship. June turned into July. When August arrived, they realized a ship was not coming.

"I have faith in our government. A ship will come next summer."

"I fear there will be no ship then either."

"I fear this too."

The *Neptune* had tried to reach the expedition team. The ship had left St John's on 8 July 1882.

But ice filled the waters. This forced the ship to stop at Littleton Island and Cape Sabine, 250 miles away from Fort Conger. The crew unloaded a small food supply.

"Just leave a few crates. They have plenty of animals to hunt!"

10

A dangerous journey

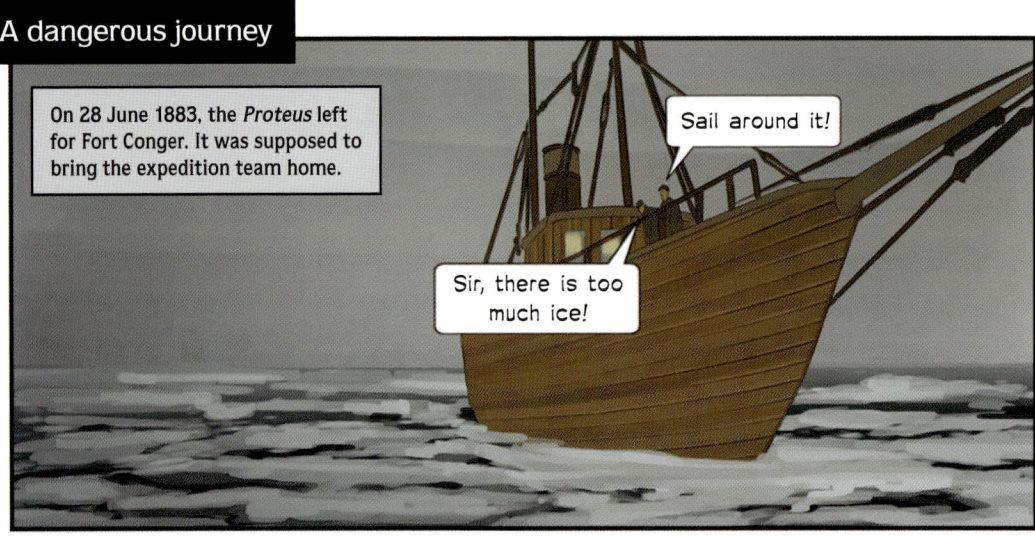

The beginning of the end

There was no rescue party in sight on Cape Sabine. Instead, Greely and his men found a note from the failed supply mission.

"The *Proteus* sank. They have left us three crates of food."

"That will last us only a few weeks!"

Greely still hoped rescuers would come. In any case, the men could not make the trip back to Fort Conger. So, the crew settled in. They built a hut called Camp Clay.

Greely regained focus. He prepared his men to face a third Arctic winter. Many were now glad of his leadership.

"There are few animals to hunt here. We must ration food."

"Yes, sir."

"Yes, sir."

"Yes, sir."

Still, the men struggled. Many had frostbite. Their food supplies were extremely low. In March, one man was caught taking food. Greely warned that anyone found stealing again would be shot. At the start of April, two men died of starvation.

One by one, death came for the men. Within hours of Frederick's return, Lockwood died. Three days later, another crew member passed away. Both died from starvation and the intense cold.

In late April, two men went seal hunting. One fell from a kayak and drowned.

Edward! Watch out!

Ahh!

In early June, Charles Henry was shot for stealing food again.

You were warned!

While Greely and his men suffered, Henrietta Greely was trying to help them. After the *Proteus* sank, she wrote to the government. She asked them to send another rescue ship. Her letters were ignored.

Mr Greely expressed to me complete faith in the government's care for its own expedition.

Many officials believed the expedition team was fine.

Hunting and rationing should keep the men comfortable through the winter.

We can send a ship next spring.

They have abandoned those poor men!

Henrietta went to the press. Embarrassed by the newspaper stories, the government agreed to act. In April, three rescue ships set sail.

On 22 June 1884, rescuers reached Cape Sabine. They had expected to find 25 healthy men.

Instead, they found seven sickly survivors. Greely and his crew had been stuck there for nine months.

Greely, is this you?

The rescuers gathered the team's scientific records. They exhumed the bodies of the dead. The expedition members, both dead and alive, were finally going home.

The world would learn about the sacrifices and achievements of the Greely expedition.

MAP OF THE EXPEDITION

The Greely expedition builds Fort Conger, their headquarters for the two-year mission, on Ellesmere Island.

Destination: Lady Franklin Bay

Fort Conger

Crew members Lockwood, Brainard and Christiansen break the British record for Furthest North at 83 degrees, 23 minutes, 8 seconds North.

Furthest North

The first supply ship, the *Neptune*, leaves a small food supply here for the Greely expedition. It returns home because ice blocks the way forward. Later, Greely's crew sails from Fort Conger to Cape Sabine. They build a second camp here called Camp Clay. They are eventually found by the rescue crew.

Cape Sabine

Littleton Island

Ellesmere Island

Smith Sound

Greenland

The second supply ship, *Proteus*, is trapped by ice near Smith Sound. It sinks.

***Neptune* also unloads some supplies at Littleton. The *Proteus* crew was supposed to wait for the Greely expedition on this island, but they return home after their rescue.**

Canada

Departure: St John's, Newfoundland, Canada

MORE ABOUT THE EXPEDITION

The polar expedition survivors faced difficult times after their rescue. Upset by the cannibalism rumours and Greely's decisions, the army ignored the men's scientific research. They did not pay the men's salaries for years. Slowly, the army began to recognize the achievements of the survivors. Each man was given a promotion. Greely refused one.

Little by little, the survivors returned to normal life. Maurice Connell, Julius Frederick and Henry Biederbick all went on to work for the US Weather Bureau. David Brainard stayed in the army. Eventually, he wrote two books about the journey to Lady Franklin Bay. Francis Long went on another polar expedition.

Adolphus Greely stayed in the army until he retired. Over time, Greely gained praise for his leadership in the Arctic. He received the Congressional Medal of Honor in 1935. It was just months before his death.

The expedition's research had been part of a larger effort called the First International Polar Year. The work of Greely and his men plays an important role in modern-day research. Scientists of today can look back at the weather data from nearly a century ago. They can compare it to recent data. The research of the Greely expedition is helping scientists understand how Earth's climate is changing.

GLOSSARY

barren growing little to no plants

cannibalism act of a person eating the flesh of another person

exhume take a dead body out of the place it was buried

expedition journey with a goal, such as exploring or searching for something

frostbite freezing of skin on some part of the body due to cold temperatures

funding money that is given to a group or person and is to be used for a specific purpose

Inuit native people of Alaska and other northern and Arctic regions

latitude distance north or south from the equator, measured in degrees

ration give out something in small amounts in order to keep it from running out

scurvy deadly disease caused by lack of vitamin C; scurvy causes swollen limbs, bleeding gums and weakness

telegraph electric device or system for sending messages by a code over wires

US Congress law-making body of the United States

FIND OUT MORE

BOOKS

Explorers (Collins Fascinating Facts) (Collins, 2016)

Explorers: Amazing Tales of the World's Greatest Adventurers, Nellie Huang (DK Children, 2019)

Explorers of the Coldest Places on Earth (Extreme Explorers), Nel Yomtov (Raintree, 2021).

WEBSITES

www.dkfindout.com/uk/history/explorers/polar-explorers/
Read more about explorers with DKFindout!

www.natgeokids.com/uk/discover/geography/general-geography/ten-facts-about-the-arctic/
Learn some interesting facts about the Arctic with National Geographic.

AUTHOR BIO

Golriz Golkar is the author of more than 60 non-fiction and fiction books for children. Inspired by her work as a primary school teacher, she loves to write the kinds of books that students are excited to read. Golriz holds a degree in American literature and Culture from UCLA and Master of Education in language and literacy from the Harvard Graduate School of Education. Golriz lives in France with her husband and young daughter.

ILLUSTRATOR BIO

Ana Carolina Tega is a Brazilian digital artist. She is an undergraduate student from the University of Hertfordshire and is currently studying 3D games art and design. She also works in the entertainment industry for short film production. Her body of work includes developing 3D assets for games, creating concept art for films and illustrating for the publishing market. Today, she is part of the Storm Creative Studio artist team.